FLAMINGOS

FLAMINGOS

ODYSSEYS

MELISSA GISH

CREATIVE EDUCATION · CREATIVE PAPERBACKS

Published by Creative Education and Creative Paperbacks
P.O. Box 227, Mankato, Minnesota 56002
Creative Education and Creative Paperbacks
are imprints of The Creative Company
www.thecreativecompany.us

Design by Blue Design (Bluedes.com)
Art direction by Wyeth Morgan

Images by Alamy Stock Photo/blickwinkel/M. Woike, 50, Claudio Contreras, 28, Lebrecht Music and Arts Photo Library, 54; Dreamstime/Attila Tatár, 12, Galyna Andrushko, 27, Jakezc, 64; Getty Images/Geng Zhang / 500px, 35, imageBROKER/Ronald Wittek, 69, James Warwick, 11, Manoj Shah, 42, © Marco Bottigelli, 63, Vicki Jauron, Babylon and Beyond Photography, 40; iStock/FOTOGRAFIA INC., 57, Overlook, 22, Simon Podgorsek, 66; Shutterstock/Steffen Foerster, 2; SuperStock/Elliott Neep/ Minden Pictures, 32–33, John Warburton Lee, 37; Unsplash/ Alejandro Contreras, cover, Bibhash (Polygon.Cafe) Banerjee, 8, Leandra Rieger, 4–5, Lieselot., 24; Wikimedia Commons/Bukvoed, 46, Carlos Urzua, 70–71, Foncea, 17, John Singer Sargent, 49, Luca Galuzzi, 75, Mark Catesby, 60, Pedro Szekely, 18, public domain, 58–59, Robert Claypool, 6

Library of Congress Cataloging-in-Publication Data
Library of Congress Cataloging-in-Publication Data
Names: Gish, Melissa, author.
Title: Flamingos / Melissa Gish.
Other titles: Flamingos (Odysseys in the wild)
Description: Mankato, Minnesota : Creative Education and Creative Paperbacks, [2026] | Series: Odysseys in the wild | Includes bibliographical references and index. | Audience: Ages 12–15 | Audience: Grades 7–9 | Summary: "Discover all there is to know about flamingos, the majestic pink birds, in this high-school nonfiction text exploring the lives of these wild animals"– Provided by publisher.
Identifiers: LCCN 2024045511 (print) | LCCN 2024045512 (ebook) | ISBN 9798889896562 (library binding) | ISBN 9781682778227 (paperback) | ISBN 9798889897361 (ebook)
Subjects: LCSH: Flamingos–Juvenile literature.
Classification: LCC QL696.C56 G576 2026 (print) | LCC QL696.C56 (ebook) | DDC 598.3/5-dc23/eng/20241219
LC record available at https://lccn.loc.gov/2024045511
LC ebook record available at https://lccn.loc.gov/2024045512

Printed in India

American flamingo

CONTENTS

Introduction

It is late spring at Kenya's Lake Nakuru National Park, and more than 2,000 lesser flamingos have arrived to join a colony numbering in the 100,000s. For each bird, the time has come to select a mate, and with tens of thousands of individuals from which to choose, the choice is neither quick nor easy. Small groups form as the birds begin their mating dance by scurrying through the shallow

OPPOSITE: A large group of flamingos is called a flamboyance.

water at the lake's edge. Moving as one, they turn their heads to the left and then to the right—repeating this motion as they zigzag through the water like a floating carpet of pink feathers. A female bird breaks away from the crowd. An interested male follows her. The two circle each other, bobbing their heads to signal their acceptance of the other. They will remain bonded with each other as they build a nest and raise a chick.

Some flamingos have been observed to mate for life.

Meet the Flamingo

Flamingos are some of the most remarkable birds in the animal kingdom. They have the longest legs and longest necks in proportion to their body size of any bird, and they are among the most colorful large birds on the planet.

Flamingos are the sole members of the genus *Phoenicopterus* (*FEE-nih-KOP-ter-uhs*). This name is Greek for "purple wing." The closest relatives of flamingos include storks, herons,

OPPOSITE: Flamingos can fly long distances at speeds of up to 35 miles (56.3 kilometers) per hour.

egrets, ibises, and spoonbills—all large wading birds. Flamingos also share many characteristics with geese, such as the webbing between their toes and waterproof feathers. Like most birds, flamingos exhibit sexual dimorphism, which means males and females differ in appearance. In the case of flamingos, males are about one-third larger than their female counterparts.

The six flamingo species are divided into two general groups, Old World and New World, according to their general geographic location. The two Old World species, the greater flamingo and the lesser flamingo, are the largest and smallest species respectively. Greater flamingos weigh up to 9 pounds (4.1 kilograms) and have a wingspan of about 5.5 feet (1.7 meters). Lesser flamingos average 3.5 pounds (1.6 kg) in weight and have a wingspan of slightly more than 3 feet (0.9 m). Old World flamingos are native to coastal areas of southern Europe, Africa, the Middle East, and western India.

There are four species of New World flamingos: American (or Caribbean), Andean, Chilean, and James's flamingos. Male American flamingos are the largest of the New World flamingos. These birds average 6 pounds (2.7 kg) in weight and have a wingspan of about 5 feet (1.5 m). American flamingos inhabit the Galápagos Islands and coasts and islands from Colombia to Venezuela as well as islands in the

Caribbean Sea. Also, small flocks of American flamingos have been spotted in southern Florida's Everglades National Park—most likely they escaped from zoos or accidentally traveled to the United States from Cuba.

Andean flamingos are the only flamingos with bright yellow legs, as other flamingos typically have orange or pink coloration. They live in wetlands of the Andes Mountains in southern Peru, Bolivia, northern Chile, and northwestern Argentina. Chilean flamingos share the Andean flamingos' range but also extend into northern Peru, Ecuador, and western Brazil. Chilean flamingos have a characteristic red or pink ring around their ankle joints about halfway up the leg. (A flamingo's knees are actually hidden under its feathers near the top of the leg.) James's flamingos, also known as puna flamingos, were named for British naturalist Harry Berkeley James,

Chilean flamingo

New Species

British naturalist Harry Berkeley James was first a businessman who lived and traveled throughout South America. He faced immense challenges including war, the destruction of his home by a tidal wave, and the decline of his business. In 1881, James moved to Valparaíso, Chile, where he began to study birds more extensively. With the help of German naturalist Carlos Rhaner and British zoologist Philip Sclater, James collected and examined flamingos from across Chile, eventually discovering the species now known as the James's flamingo. James's findings are explored in his book *New List of Chilian Birds*, which was published in 1892 after his death.

who first published information on this flamingo species in 1885. Only slightly larger than the lesser flamingo, the James's flamingo is the smallest of the New World flamingos. This flamingo also shares the Andean flamingo's range.

Like other birds, flamingos are warm-blooded, feathered, beaked animals that walk on two feet and lay eggs. Flamingos are tropical water birds that spend their lives in warm, muddy lagoons, **estuaries**, coastal wetlands, and shallow lakes—particularly seasonally occurring

soda lakes. Flamingos are able to float on water and propel themselves forward by paddling with their webbed feet. However, they prefer to walk through shallow water and will typically fly over stretches of deep water rather than swim. The flamingo's long, flexible neck and spindly legs—up to 49 inches (124 centimeters) in length, depending on the species—allow this bird to feed in water that is often too deep for other kinds of water birds. All flamingos have three forward-pointing toes, but the greater, lesser, American, and Chilean flamingos also have one backward-pointing toe called a hallux. The flamingo stomps its webbed feet to stir up water, loosening food from the muddy bottom.

Flamingos feed with their heads upside down, holding their breath when submerged underwater. The flamingo's

"FLAMINGOS FEED WITH THEIR HEADS UPSIDE DOWN, HOLDING THEIR BREATH WHEN SUBMERGED UNDERWATER."

bowl-shaped bill, which varies in its coloration pattern from species to species, is made of keratin, the same hard but flexible substance found in human fingernails. The lower jaw is attached to the flamingo's skull, while the upper jaw performs all the movement—this is unlike the jaw structure of most other birds. Flamingos feed on

OPPOSITE Flamingos have 19 neck bones, called cervical vertebrae, which allow the neck to twist and bend.

plant seeds as well as a variety of living creatures—algae, tiny fish, and invertebrates—by sifting these organisms from the water.

Rows of filtering plates are located on the edges of the upper and lower jaws. These plates, called lamellae, are shaped like combs with frayed edges, and they work like sieves. As the flamingo scoops up a mouthful of water, it loosely closes its bill. It then presses its fat, bristle-edged tongue against the roof of its mouth to force out the water through the lamellae, trapping the

Balancing Act

Along with their pink feathers and long necks, flamingos are known for balancing on one leg. There are two main theories for why flamingos do this. The first is that it keeps them warm. Research has shown that flamingos are more likely to balance on one leg in colder weather. This leads scientists to believe that by tucking one leg into their feathers, flamingos are able to conserve body heat. The other theory is that standing on one leg requires less effort. A flamingo's legs can be locked into position, which means it uses less muscle to balance on one leg than to support itself with two.

organisms inside its mouth. The flamingo performs this filtering action with great speed as it swings its head to and fro through the water or mud.

Greater, American, and Chilean flamingos have widely spaced filters for capturing seeds and prey such as worms, fish, insects, and mollusks up to 1 inch (2.5 cm) long. Pressing their tongue against the roof of their mouth 4 to 5 times per second, they consume roughly 9.5 ounces (269 grams) of food per day. Lesser, Andean, and James's flamingos have narrowly spaced filters designed to catch tiny flies, shrimp, and single-celled organisms such as algae and plankton. Their tongues move 20 times per second, filtering a little more than 2 ounces (56.7 g) of food per day.

The foods that flamingos eat are what gives these birds their brilliant color. Algae and bacteria contain **pigments** called carotenoids, which give color to plants and animals. When flamingos eat carotenoid-producing organisms or other creatures, such as shrimp, that feed on these smaller organisms, the flamingos themselves become darker in color. Lesser flamingos, for example, are pale pink for most of the year, but when they feed on large amounts of carotenoid-rich algae during breeding season, their feathers, legs, and even eyes turn a brilliant crimson. Captive flamingos whose diets do not consist of carotenoid-rich foods will become paler in color, which is why most zoos today supplement flamingo diets with red algae. In the wild, the deep crimson American flamingos are the most colorful, while greater flamingos are naturally the palest pink.

Salar de Uyuni, Bolivia

Flock Life

Because flamingos are lightweight birds that spend most of their time on the ground, they are vulnerable to predators. To help protect themselves, they live in flocks that may number from several hundred to several thousand. Flamingos inhabit places that most other animals find inhospitable because of the harsh extremes. They visit mangrove

OPPOSITE: Flamingos seek out lakes in which few or no fish live because fish compete for the flamingos' favorite food: algae.

"FLAMINGO FLOCKS JOIN TOGETHER TO FORM ENORMOUS COLONIES AT SODA LAKES IN SOUTH AMERICA OR AFRICA."

swamps, estuaries, and intertidal zones (ocean shore areas that are underwater at high tide but exposed at low tide). The water in these habitats may be very fresh if rainfall is consistent or very salty if it is infrequent. Nevertheless, such areas are rich in the **nutrients** that tiny organisms need to flourish. The most productive aquatic environments on Earth are soda lakes, which explode seasonally with protein-rich algae, insects, brine shrimp, and other tiny creatures brought on by heavy spring rains.

Flamingos do not **migrate** seasonally, but they typically move from place to place throughout the year in search of food and nest sites. Flamingo flocks join together to form enormous colonies at soda lakes in South America or Africa, where they feast in preparation for the task of breeding and raising chicks. In the heat of summer, as water **evaporates** from a soda lake, a thick crust of minerals is left to bake in the sun, forming a hard salt flat. The world's largest salt flat is Salar de Uyuni in Bolivia. This area of approximately 4,085 square miles (10,580 square kilometers) is situated in the Andes Mountains at about 12,000 feet (3,658 m) above sea level. Tens of thousands of flamingos go there to breed and raise chicks during the summer, returning to warmer, lower-altitude habitats

Flamingos need to get a running start of several strides to lift their bodies off the ground for flight.

in winter. In Africa, flamingos gather at six or seven different soda lakes in the Great Rift Valley. Northern Tanzania's Lake Natron attracts nearly 2 million lesser flamingos to its salt islands, which form in the center of water so rich in algae containing red pigments that the water appears redder than flamingo feathers.

Flamingos of all ages gather together during breeding season, but they do not begin to reproduce until they are about six years old. If food is not abundant, flamingos may not reproduce at all in a given year, but if conditions

Flamingo performing a courtship display

are favorable, breeding occurs among most eligible members of a colony. Flamingos attempt to impress potential mates by participating in an assortment of synchronized movements called courtship displays. The type of display varies by species but includes activities such as marching, head flagging, wing saluting, and leg-wing stretching, and all are performed in unison. When marching, flamingos bunch together as closely as possible and move in one direction; then they suddenly turn and move in another direction. While grouped tightly together, flamingos perform head flagging by stretching their necks and lifting their bills upward as high as they can. They then swing their heads from the right to the left and back again in rapid harmony. Wing saluting is performed by repeatedly stretching the neck forward, raising the tail, and outstretching the wings for a few seconds. Flamingos also extend one leg straight back at the same time as they stretch one wing straight back in leg-wing stretching.

Flamingos form pair bonds that last through a breeding season. Together, the two flamingos use their bills to scrape up a tower of salt and mud with a shallow bowl in the center. The sun bakes this structure, which can stand as high as 24 inches (61 cm), to rock-hard strength.

Beautiful and Dangerous

Lake Natron, Tanzania, is home to the largest breeding ground of lesser flamingos in the world. However, the waters are deadly to most other animals. The same algae that give the lake its unique red color can also be poisonous when eaten. While flamingos consume the algae with no problems, the algae cause damage to the cells and organs of other birds and animals, which is why so few other species survive in this habitat. Lake Natron is also extremely hot and acidic. Flamingos have tough skin that protects them, but the waters can easily burn other animals and even humans.

Pairs spend up to six weeks bonding and nest building. Then a single chalky-white egg, about twice as big as a chicken egg, is laid. The tower protects the egg from a flood, which sometimes occurs if the salt flat cracks, and keeps the egg cooler than if it were laid directly on the ground. The parents take turns sitting on or standing over the egg, incubating it for 27 to 31 days and gently turning it daily.

Using its **egg tooth**, the chick chips through the hard shell of its egg. This may take between 24 and 36 hours, during which time the chick cheeps constantly, and the parents answer in order to establish a voice bond that will allow them to recognize each other later. The parents will feed only their own chick. Newly hatched chicks weigh between two and three ounces (56.7–85 g). They are covered with fluffy gray or white **down**, and their bills are straight and red. Chicks are weak and fragile, but they can lift their heads long enough to be fed crop milk by their parents. Crop milk is a red liquid secreted by **glands** in the adult flamingo's throat. It is 9 percent protein and 15 percent fat, and it contains some of the adult's own red and white blood cells. Chicks eat crop milk for two months as the filter system in their bills develops.

eek-old chicks wander from their nests and gather in groups called crèches. Here they find relative safety in numbers.

Marabou Stork: Family or Foe?

As members of the stork family, marabou storks are flamingo relatives. Yet comparing a flamingo to a marabou is like comparing Cinderella to her ugly stepsisters. Flamingos are beloved by many for their colorful feathers, long necks and legs, and elegant movements. On the other hand, marabou storks are characterized by their shaggy black coats, massive beaks, and bald heads, making them the stuff of nightmares. Another distinctive feature of the marabou is a pink pouch on its throat, which is believed to be used for courtship displays rather than storing food. The pouch is also used to make noises such as grunts and croaks.

At about 11 weeks old, chicks start to grow flight feathers, and their bills begin to curve. Juvenile flamingos remain gray or white until they reach two or three years of age, when their feathers turn pink.

Both adult and young flamingos are threatened by a variety of predators. However, chicks and eggs are particularly vulnerable, since flamingos do not have the body strength or bill design to fend off attackers and protect their young. In South America and the Bahamas, foxes, wild cats, eagles, hawks, **feral** pigs, and snakes attack flamingos and raid their nests. Leopards, cheetahs,

Marabou storks are found south of the Sahara Desert in Africa.

jackals, and baboons regularly attack adult Old World flamingos, while hyenas, warthogs, mongooses, and wild dogs typically target eggs and chicks.

More threatening to Old World flamingos than any of these other predators, though, is the marabou stork, also called the undertaker stork because of its long, black coat of feathers and nearly bald head. Just a few of these 20-pound (9.1-kg) birds with their thick, sharp, 14-inch-long (36-cm) bills can kill hundreds of flamingo chicks in a matter of hours. As devastating as this may seem, in a typical colony of 500,000 lesser flamingos on an African lake, the **mortality rate** of chicks may be as low as 5 percent.

Flamingos and People

The flamingo has been part of the **mythology** and legends of many **cultures** for thousands of years. In ancient Egypt, the flamingo signified the color red, and Egyptian mythology explained that when a holy tree burst into flames, the sun god Ra emerged from the fire as the Bennu, a brilliant crimson-colored bird. The flamingo, which once lived in great numbers along the Nile River, became the living embodiment of the Bennu, and this bird was believed to carry the soul of Ra. In the town of Naqada, on the western bank of the Nile, archaeologists discovered images of flamingos painted on pottery that is more than 3,200 years old.

The story of the Bennu perhaps gave rise to the Greek myth of the phoenix in the fifth century B.C. Every 1,000 years, this bird would burn itself and be reborn from the ashes. Similar mythical birds can be found in symbolic tales of transformation the world over, from the Iranian huma to the Slavic firebird.

When the Romans visited Egypt in the first century B.C., they captured flamingos and served the birds' tongues to royalty. In the North African country of Tunisia, flamingo meat was still considered a delicacy hundreds

Birds Mosaic, Caesarea, Isreal

of years later. At an archaeological site in El-Djem that dates to the third century A.D., **mosaic** artwork was found depicting a flamingo being prepared for cooking. Around the same time, flamingos were also kept as pets and displayed in traveling circuses and private zoos. On the island of Sicily, another mosaic was found—this one dating to the fourth century A.D.—of a child's chariot being pulled by red flamingos.

In the Americas, the Moche civilization arose about 2,000 years ago in northwestern Peru. The Moche often depicted animals and birds—including flamingos—in their art and pottery. In 1987, Peruvian archaeologist Walter Alva discovered a tomb of a Moche king dating to the second century A.D. Among the artifacts found there were ornaments and headdresses made of flamingo feathers.

"WHEN THE ROMANS VISITED EGYPT IN THE FIRST CENTURY B.C., THEY CAPTURED FLAMINGOS AND SERVED THE BIRDS' TONGUES TO ROYALTY."

While flamingos are no longer widely worshiped as gods or magical beasts, they are still greatly admired for their beauty and grace. Flamingos are a mainstay of zoos around the world, but an Australian ban on exotic bird importation left only two specimens as the sole captive flamingos on the entire continent—and they also happened to be the oldest flamingos in the world. A greater

Dance of the Flamingos

Flamenco is the Spanish word for "flamingo." It is also the name of a famous art form that originated in southern Spain. Flamenco, which is often referred to as flamenco dancing, is a traditional performance dating back to the 15th century. It combines elements of song, dance, and instrumental music. The reason why this art form shares a name with flamingos is not entirely known, but some believe the dance is meant to represent the quick movements that flamingos make when performing courtship displays. Flamenco costumes are also traditionally a bright red color with many ruffles, possibly resembling a flamingo's feathers.

The estimated lifespan of a wild flamingo is 20 to 30 years.

flamingo, nicknamed "Greater," arrived at the Adelaide Zoo in 1933 and is thought to have lived more than 80 years. A Chilean flamingo arrived in 1948, just prior to the ban. The two flamingos displayed no fear of people despite an incident in 2008, when four teenagers attacked Greater, fracturing his skull and blinding him in one eye. After a year of medical treatments and therapy, Greater recovered and returned to his zoo home, taking up his familiar post beside his enclosure partner of more than 50 years. Greater died in 2014.

In the United States, flamingos have often been viewed as comedic, showy characters. People who share such traits as long legs are sometimes associated with the birds. Infamous New York gangster Benjamin "Bugsy" Siegel had a girlfriend named Virginia Hill, who had long, skinny legs, so Siegel called her "Flamingo." In

1945, Siegel went to Las Vegas to build a casino—the biggest, fanciest casino anyone had ever built at the time. He spent $6 million on the lavish building—and he named it the Flamingo in Virginia's honor. Today, the Flamingo Las Vegas, which still features tropical décor and a garden courtyard that is home to a flock of live Chilean flamingos, is the oldest casino in operation on the Las Vegas Strip.

A troupe of comical flamingos with super-skinny legs appears in the Disney film *Fantasia 2000*. The birds struggle to keep one member of the troupe in step, but he's too busy playing with a yo-yo—and entangling his fellow flamingos—to follow his dance routine. Known as Yo-Yo Flamingo, the character also appeared in several episodes of *House of Mouse*, a Disney television series from the early 2000s. Another flamingo entertainer was Placido Flamingo, a *Sesame Street* character modeled after Spanish opera singer Plácido Domingo. The singing bird appeared at the Nestopolitan Opera, a fictional venue created to introduce classical music to *Sesame Street* viewers. Real flamingos are the stars of Disneynature's 2008 film *The Crimson Wing: Mystery of the Flamingos*, which tells the story of a colony of nesting flamingos on a salt island in the Great Rift Valley's Lake Natron.

In fictional books, flamingo characters often find themselves standing out from other kinds of birds. Jill

OPPOSITE A game of croquet in the book *Alice in Wonderland* features flamingos as mallets.

Ker Conway's 2006 book *Felipe the Flamingo* tells of a young flamingo who is left behind when his flock migrates. Other birds adopt him until his family returns. And in *Sylvie* (2009), by Jennifer Sattler, a young flamingo begins experimenting with foods that change her pink coloration. Jamie Harper's book *Miss Mingo and the First Day of School* (2009) presents a flamingo teacher getting to know her students.

Perhaps the most familiar flamingos of all are the plastic lawn flamingos. First designed in 1957 by Don Featherstone for a Massachusetts plastics company, the pink lawn ornaments have become so widely known that they have inspired characters in film. In the 2011 animated movie *Gnomeo and Juliet*, two love-struck lawn gnomes make friends with a lonely plastic flamingo named Featherstone. In 1987, "Pink Floyd" the Chilean

"THE MOST FAMILIAR FLAMINGOS OF ALL ARE THE PLASTIC LAWN FLAMINGOS."

flamingo escaped from Tracy Aviary in Salt Lake City, Utah, and took up residence at the Great Salt Lake. City officials planted a flock of plastic pink flamingos around the lake to help Pink Floyd feel less lonely, which appeared to work, since the bird continued to live at the lake until 2005.

Since 2002, original Featherstone-designed lawn flamingos have been available online through GetFlocked.com. In 2004, the company sent a box of flamingos to American soldiers in Iraq to remind them of their homes

Plastic lawn flamingo

and neighborhoods. The soldiers wrote back, expressing their delight at the surprise, saying, "your birds have brought nothing but smiles to the people who see them... They are pure Americana."

A Flamingo's Paradise

After escaping from Tracy Aviary in Salt Lake City, Utah, the now-famous Chilean flamingo Pink Floyd made his home at the Great Salt Lake, which lies just west of the city. He spent every winter there for 17 years. Before arriving in the United States, Pink Floyd was believed to have lived at a high-altitude salt lake in Chile, which explains why the waters of the Great Salt Lake appealed to the bird. Not only does the lake have a high salt concentration, but it is also home to brine shrimp, a popular food source for Chilean flamingos. No wonder Pink Floyd felt right at home!

Research and Conservation

The first flamingo ancestor most closely related to modern flamingos was *Phoenicopterus croizeti*, which lived about 40 million years ago. *Phoenicopterus croizeti* was one of the first birds to wade and feed in shallow water but, unlike modern flamingos, was ill-equipped for distance flying. Fossils of it have been found in France. Another flamingo ancestor shared prehistoric wetland and rainforest

OPPOSITE: Early flamingo ancestors are believed to have had shorter wings and bills and to have spent more time in the water than modern-day flamingos.

habitats of central Australia with many other bird species as well as early crocodiles about 25 million years ago. Standing about 5 feet (1.5 m) tall, *Phoeniconotius eyrensis* was one of the largest flamingo ancestors ever discovered. Flamingos disappeared from Australia millions of years ago as climate change caused much of their habitat to dry up and turn to desert. Many fossils of early flamingos and other wading birds have been found in the province of South Australia at Lake Eyre, a shallow salt flat with sparse seasonal water.

On the other side of the world, one of the smallest and most recent flamingo ancestors existed in North America until the last glacial period ended about 11,000 years ago. Fossilized remains of *Phoenicopterus minutus*, whose name means "little flamingo," were first found in California's Mojave Desert in the 1950s. Long ago,

Flamingos on the sand dunes of Walvis Bay, Namibia

OPPOSITE The flamingo's 12 principal flight feathers, which are black, are visible only when the wings are extended.

North America's flamingos were forced to move toward the equator, where they continued to **evolve** into the modern species living on Earth today.

Flamingos are uniquely suited to their shallow-water environments, which is why scientists and conservationists are concerned about these birds' future on the planet. Flamingo habitats are being disturbed and even transformed by industry, particularly by the mining of sodium carbonate, the salty mineral present in the very lakes where flamingos feed and nest. Dozens of studies are being conducted in countries containing flamingo habitats—including Afghanistan, Argentina, India, Kenya, Siberia, Tanzania, and Turkey—to determine how mining might affect flamingo breeding and survival.

Sodium carbonate is used in a variety of products, from glass to detergent to toothpaste, and it is even used

as a food additive to keep powdery foods such as cocoa mix and dry oatmeal from caking. To extract this chemical from soda lakes, water is pumped into enclosed pits and allowed to evaporate. The remaining salty residue hardens in the sun and is then cut into blocks. Taking water from the lakes in this way speeds up the lakes' own evaporation process, leaving flamingos with less water for feeding and fewer salt islands for nesting and raising young. In addition, the human activities disturb flamingos, who may then avoid breeding grounds altogether.

Scientists are particularly alarmed at the Tanzanian government's approval of a major soda mining operation on Lake Natron, where 65 to 75 percent of the world's lesser flamingo population nests each year. Continued mining in Africa could not only decimate flamingo populations, but it could also affect the lives of the people in mining nations. For example, thousands of people visit Lake Natron each year to see the flamingos, providing the Tanzanian people with tourism dollars. While soda mining may be profitable in the short term, the soda supply will not last forever. Without flamingos to draw tourists year after year, the Tanzanian economy will suffer.

Industrialists believe the populations of roughly 2.2 to 3.2 million lesser flamingos and 550,000 greater

OPPOSITE Stress caused by human interference can lead flamingos to abandon their unhatched offspring.

flamingos are stable, yet flamingos are part of a delicate ecosystem that is being altered, leading to a persistent annual decline in flamingo numbers. Populations of New World flamingos, which are already smaller (with James's flamingos numbering about 64,000 and Andean flamingos fewer than 34,000), face similar challenges. In South America, mining operations and crop irrigation deplete water from many flamingo habitats. As cities expand, the building of roads and pollution from sewage also contribute to habitat destruction. In addition, thousands of flamingo eggs are collected—even from protected areas—to be illegally sold for food, and flamingos can fall victim to **poaching** for their oil and feathers, materials that many **indigenous** people believe can prevent and cure a variety of health problems. The population of roughly 200,000 Chilean flamingos is affected by tourists and

Flamingo Hotspot

Los Flamencos National Reserve is located in the Atacama Desert of northern Chile. It spans 182,823 acres (73,986 hectares) and is divided into seven regions, each with its own landscape and ecosystem. These range from deserts and hills to lagoons and even volcanoes. However, one thing the regions share are the flamingos. The reserve is home to all three flamingo species found in Chile: Chilean flamingos, Andean flamingos, and James's flamingos. The Tara and Pujsa salt flats are two of the most popular flamingo hangouts—and popular destinations for tourists—but flamingos also inhabit the Tamarugo forest and the Miscanti and Miñiques lagoons.

photographers as well, whose interference often leads to nest abandonment.

In North America, decades of overhunting and habitat destruction in the early 20th century led to a sharp decline in American flamingo populations, and by the mid-20th century, fewer than 22,000 of these birds remained. However, conservation efforts, including the establishment of wildlife reserves, enabled the flamingos to mount a strong comeback. The population of American flamingos today is estimated to be roughly 850,000.

Although their numbers continue to decline, government-sponsored protected areas offer some support for the other three species of New World flamingos. One of the most valuable flamingo habitats in South America is the Eduardo Avaroa Andean Fauna National Reserve in southwestern Bolivia, which covers more than 1.7 million acres (687,966 ha). The Salinas and Aguada Blanca Natural Reserve in Peru also provides protected nesting grounds high in the Andes Mountains. Los Flamencos National Reserve in Chile is another important flamingo habitat. The reserve's major lake, Salar de Tara (Tara Salt Flat) was designated a Wetland of International Importance in 1996 by the Ramsar Convention, an international treaty for the conservation of wetlands. In addition to protecting flamingos, the site also offers sanctuary to many of the flamingo's high-altitude neighbors, including the Andean goose; the tall, flightless Darwin's rhea; and the quail-like puna tinamou.

Park rangers patrol flamingo nest sites to offer protection from poachers, and researchers continue to monitor flamingo populations. As of 2013, Argentina, Bolivia, and Chile were working to create a tri-national reserve to provide even more protected flamingo habitat. While such measures are positive developments, both human interference and climate change persist in affecting flamingo behavior and mortality around the world. Flamingos have **adapted** to their harsh environments over countless generations. Forcing these birds to adapt further to even more challenging conditions as a result of human activities seriously imperils their ability to survive and thrive. Conducting further research and strengthening conservation measures will be necessary steps to take to save flamingos from an uncertain future.

Rare Species

Andean flamingos are some of the rarest flamingos in the world, found only in the Andes Mountains of South America. They rely on wetlands for breeding and food, yet these habitats are becoming increasingly hard to find. Rising global temperatures cause wetlands to dry up. Human activity, such as lithium mining in the Andes, pollutes and even destroys wetland habitats for flamingos and other wildlife. While Andean flamingos are currently not endangered, threats such as climate change and habitat loss could put them at risk of extinction in the near future.

Selected Bibliography

Aeberhard, Matthew, and Leander Ward. *The Crimson Wing: Mystery of the Flamingos*. DVD. Paris, France: Disneynature, 2008.

Collar, Nigel. *Pink Flamingos*. New York: Abbeville Press, 2000.

Flamingo Resource Centre. "Flamingo Basics." http://www.flamingoresources.org/basics.html.

McMillan, Bruce. *Wild Flamingos*. Boston: Houghton Mifflin, 1997.

San Diego Zoo. "San Diego Zoo Animals: Flamingo." http://animals.sandiegozoo.org/animals/flamingo.

SeaWorld Education Department. "Flamingos." http://www.seaworld.org/animal-info/info-books/flamingo/pdf/ib-flamingo.pdf. SeaWorld, 2005.

Glossary

adapt to change to improve the chances of survival in an environment

archaeologist a person who studies human history by examining ancient peoples and their artifacts

culture a particular group in a society that share behaviors and characteristics that are accepted as normal by that group

down a layer of soft feathers next to a bird's skin that traps in heat to keep the bird warm

egg tooth a hard, toothlike tip of a young bird's beak or a young reptile's mouth, used only for breaking through its egg

estuary the mouth of a large river, where the tides (from oceans or seas) meet the streams

evaporate to change from liquid to invisible vapor, or gas

evolve to gradually develop into a new form

feral in a wild state after having been domesticated

gland an organ in a human or animal body that produces chemical substances used by other parts of the body

indigenous originating in or native to a particular region or country

invertebrate an animal that does not have a backbone

migrate to travel from one region or climate to another for feeding or breeding purposes

mortality rate the number of deaths in a certain area or period

mosaic a picture or design made by arranging small pieces of colored material such as glass, stone, or tile

mythology a collection of myths, or popular, traditional beliefs or stories that explain how something came to be or that are associated with a person or object

nutrient a substance that gives an animal energy and helps it grow

pigment a material or substance present in the tissues of animals or plants that gives them their natural coloring

poaching hunting protected species of wild animals, even though doing so is against the law

soda lake a lake characterized by a heavy concentration of salt and related chemicals

Websites

Flamingo

https://www.birdlife.org/birds/flamingo/

Learn more about the different flamingo species and recent conservation efforts.

Lake Natron

https://www.brilliant-africa.com/tanzania/lake-natron

Read about the flamingos and other wildlife that inhabit Lake Natron, Tanzania.

These Flamingos Have Sweet Dance Moves

https://www.youtube.com/watch?v=QLV_K7DVeyU

Watch the dance-like courtship displays of flamingos in Laguna Brava, Argentina.

Index